REVOLUTIONARY DISCOVERIES OF SCIENTIFIC PIONEERS™

THE PERIODIC TABLE OF ELEMENTS AND DMITRY MENDELEYEV

FRED BORTZ

ROSEN PUBLISHING®

New York

Published in 2014 by The Rosen Publishing Group, Inc.
29 East 21st Street, New York, NY 10010

First Edition

Library of Congress Cataloging-in-Publication Data

Bortz, Fred, 1944-
The periodic table of elements and Dmitry Mendeleyev/Fred Bortz.
    pages cm.—(Revolutionary discoveries of scientific pioneers)
Audience: Grades 7-12.
Includes bibliographical references and index.
ISBN 978-1-4777-1807-0 (library binding)
1. Mendeleyev, Dmitry Ivanovich, 1834-1907. 2. Chemists—Russia (Federation)—Biography. 3. Periodic law. 4. Periodic law—Tables. I. Title.
QD22.M43B67 2014
540.92—dc23
[B]
                                                                      2013012047

Manufactured in the United States of America

CPSIA Compliance Information: Batch #W14YA: For further information, contact Rosen Publishing, New York, New York, at 1-800-237-9932.

A portion of the material in this book has been derived from *Mendeleyev and the Periodic Table* by Katherine White.

CONTENTS

INTRODUCTION

In most high school or college chemistry classrooms, and in many physics classrooms as well, you will see a chart on the wall called the periodic table of the elements. It is a series of boxes arranged in rows and columns. Each box contains a one- or two-letter atomic symbol representing one of the known chemical elements—substances that have only one kind of atom—and its atomic number. The atomic numbers begin with one and increase in numerical order from left to right and top to bottom.

A second number in each box is the element's atomic weight, which is a decimal number that generally increases from one box to the next, usually by more than one unit. The boxes usually contain other information about the element they describe, such as whether the element is a solid, liquid, or gas at room temperature and normal atmospheric pressure, its melting and boiling points, its density, and other properties. Some periodic tables even include a code in each box that tells something about the internal structure of the atoms of that box's element.

The periodic table is an amazing collection of knowledge about the atoms that make up all normal matter. When you realize everything it contains, you are likely to think that it took a genius to come up with the idea—and you would be correct. The creator

WHEN RUSSIAN CHEMIST DMITRY IVANOVICH MENDELEYEV WAS WRITING A TEXTBOOK IN 1869, HE STRUGGLED WITH THE BEST WAY TO DESCRIBE ALL OF THE NEW CHEMICAL ELEMENTS THAT WERE BEING DISCOVERED. IN WHAT SEEMED LIKE A FLASH OF INSPIRATION BUT WAS ACTUALLY THE PRODUCT OF LONG AND CONTINUING STUDY AND RESEARCH, HE DEVELOPED THE FIRST VERSION OF THE PERIODIC TABLE OF THE ELEMENTS. THAT TABLE LAUNCHED A SCIENTIFIC REVOLUTION THAT CONTINUES TODAY AND STILL SERVES AS AN IMPORTANT GUIDE FOR MODERN CHEMISTRY INSTRUCTION AND RESEARCH.

of the periodic table was the great Russian scientist Dmitry Ivanovich Mendeleyev (1834–1907), who first described it in 1869.

Like most works of genius, the periodic table was the result of a combination of hard work and careful study, plus a flash of inspiration. Mendeleyev's hard work was in the field of chemistry. As nineteenth-century chemists discovered more and more elements, they began to notice that some were quite similar in their chemical and physical properties to others with quite different atomic weights, but quite different from elements with nearby atomic weights.

Mendeleyev looked for ways to arrange the various elements into sets and then to find an orderly arrangement of those sets as well. He couldn't get that problem out of his mind. He thought about it from the moment he awoke to the time he drifted off to sleep. Then one morning in 1869, he awoke with a flash of inspiration. He described it famously in this way: "I saw in a dream a table where all the elements fell into place as required. Awakening, I immediately wrote it down on a piece of paper."

Of course that flash of insight was only one step in a long and continuing process. It would not have come to him without all of his earlier study, and it would not have been recognized as a work of genius without the creative thinking and scientific work that followed. Mendeleyev's major breakthrough was his arranging of the elements in sequence by atomic weight but recognizing that there were gaps where no elements had yet been discovered.

The periodic table of the elements launched a scientific revolution that continues today, and the contents and structure of that table tell us far more than Mendeleyev could have ever imagined. This story tells the discovery and development of the periodic table and of the life of the man who first envisioned it in a dream. Read on and discover what it means to pursue a question important enough to follow for a lifetime.

CHAPTER ONE

EARLY STRUGGLES

*D*mitry Mendeleyev was born to Ivan and Mariya Mendeleyev in the town of Tobolsk in western Siberia, the vast Asian region of Russia that makes up 77 percent of the country's land area and is home to about a quarter of its population today. According to Russian custom, he was given the middle name Ivanovich, meaning son of Ivan. He was born on January 27, 1834 (according to the Julian calendar, which was in use in Russia at the time; the date would be February 8 according to the Gregorian calendar, which most of the world followed). Records from that period vary, but he was the youngest of a very large family of probably fourteen, and perhaps as many as seventeen, children.

MENDELEYEV WAS THE YOUNGEST CHILD OF A VERY LARGE FAMILY IN THE COMMERCIAL CENTER OF TOBOLSK IN WESTERN SIBERIA, SHOWN HERE IN A MODERN PHOTOGRAPH. HIS FATHER WAS DIRECTOR OF THE LOCAL GYMNASIUM, A HIGH SCHOOL THAT PREPARED STUDENTS FOR UNIVERSITY STUDIES. THE TOWN WAS ALSO HOME TO MANY GENERATIONS OF THE KORNILEVA FAMILY ON HIS MOTHER'S SIDE. HIS GRANDFATHER OWNED PAPER AND GLASS FACTORIES AND OPENED THE FIRST NEWSPAPER IN SIBERIA.

Tobolsk was an important town at the time. Located at the junction of the Tobol and Irtysh rivers, it was once a Siberian capital. The town was home to many generations of Mendeleyev's family, especially on his mother's side, the Kornilevas. The Kornileva family introduced some major businesses to the people of Siberia. Almost fifty years earlier, Mendeleyev's grandfather opened the

THE SPELLING OF RUSSIAN NAMES

The Russian language uses the Cyrillic alphabet, which is different from the Roman alphabet that people use in English. Many Cyrillic characters have equivalent letters in English, but there are some sounds that are different and have their own characters in Cyrillic. Thus Russian names can be transliterated (that is, have their letters reproduced as Roman letters) in various ways. In other works, you may see Dmitry Mendeleyev's first name spelled with an "i" at the end instead of a "y," and his last name spelled without the "y" in the final syllable. But whether you spell it Dmitri or Dmitry and Mendeleev or Mendeleyev, it is still the name of a creative revolutionary scientist.

first newspaper in the history of Siberia. He was also well known for his paper and glass factories.

MISFORTUNE STRIKES

Ivan Mendeleyev supported his large family by teaching Russian literature at the local gymnasium, where he was the director. (In Europe, a gymnasium is a secondary school that prepares students for the university.) But soon after Dmitry's birth, Ivan went blind and had to resign. Ivan's sudden blindness forced Mariya to find a way to support their large family.

YOUNG MENDELEYEV WAS NOT ENTHUSIASTIC ABOUT MANY OF HIS SCHOOL SUBJECTS, BUT HE EXCELLED IN MATH AND SCIENCE. SO HIS MOTHER INVITED HIM TO VISIT ONE OF THE KORNILEVA FAMILY GLASS FACTORIES, WHICH SHE MANAGED. HIS FASCINATION WITH THE SUBSTANCES AND PROCESSES TO MAKE GLASS WAS THE BEGINNING OF HIS INTEREST IN CHEMISTRY. THIS 1909 IMAGE SHOWING A GLASSBLOWER IN FRANCE IS NOT VERY DIFFERENT FROM WHAT DMITRY SAW AND EXPERIENCED ABOUT SIXTY-FIVE YEARS EARLIER IN SIBERIA.

Fortunately, Mariya was a smart, determined, and very open-minded woman for her time and place. In nineteenth-century Russia, most women were in charge of running the household and expected their husbands to support the family. But when family circumstances changed, Mariya knew what she had to do. She began to manage one of her father's glass factories. Some of Dmitry's first memories are of going with his mother to the Kornileva glass factory in Aremzyanka, a small village about 22 miles (35 kilometers) from Tobolsk. There he spent his days playing with the factory workers' children.

EARLY SCHOOLING

In the early years of his schooling, Dmitry was not very excited about his studies. He rarely paid attention, and his poor grades showed his lack of enthusiasm. The only subjects in which Dmitry excelled were math and science. He was especially interested in physics, the study of matter, motion, and energy. Unfortunately, Dmitry's school focused much more on classical subjects, such as famous works of literature written in Greek and Latin.

Mariya recognized Dmitry's love of science, so to support it, she invited her son to visit the glass factory as often as he liked. There he learned all about glass and glassblowing. The main ingredient in glass is common sand, which is mixed with soda and lime and heated in a furnace. Then, while it is still hot, it is shaped with special tools and air is blown into it to form bottles, glasses, or other useful objects.

Dmitry enjoyed learning about the science of glassblowing, such as how much of each material was needed for various types of glassware and how much heat was needed. Observing how all the separate ingredients, when mixed together, became an entirely new substance was the beginning of Mendeleyev's deep interest in chemistry.

Dmitry's brother-in-law, a man named Nikolai Basargin, also had a big influence on the boy's education. Basargin had been exiled to Siberia because of his role in a failed political uprising in Russia called the 1825 December Revolution, staged by a political group called the Decembrists. The group wanted Russia to change to a constitutional government and establish some form of democratic government. When the revolution failed, Russian officials sentenced the leaders of the revolution to death and banished the rest of the Decembrists to Siberia, including Basargin, who soon met and married Dmitry's older sister Olga. Noticing that Dmitry's grades were bad, Basargin began tutoring young Dmitry. Basargin was also very interested in science, so both student and teacher would read and talk about science for hours on end.

EDUCATING AN EMERGING GENIUS

By the time he turned thirteen, Dmitry's grades had greatly improved, and everyone began to notice his remarkable intelligence. Dmitry's teachers said he had a great mind for science. He would finish his school day, go home, and conduct all kinds of different scientific experiments. Soon, everyone agreed: Dmitry was brilliant.

As the youngest child, Dmitry spent an enormous amount of time with his mother. Over the years, the two developed a strong relationship. Mariya favored Dmitry because she always believed he would become a great success. When Mariya saw how much Dmitry loved science, she vowed her son would someday go on to study at a very important university.

Mariya's dream was to pay for her son's education. But it would not be easy, as the Mendeleyev family went through two sudden and deeply personal tragedies. Ivan Mendeleyev passed away in 1847. Then in 1848, the glass factory burned to the ground. The tragedies left Mariya alone and without any way to support her two youngest children. In 1849, Dmitry finished school, and in the summer of that year, Mariya gathered Dmitry and his older sister Liza and moved the family to Moscow.

The 1,300-mile (2080-km) journey was long and hard for a widow more than fifty years old, but Mariya was convinced that Moscow offered better opportunities for

TO MAKE SURE HER SON WOULD GET A STRONG EDUCATION IN SCIENCE, MENDELEYEV'S MOTHER MOVED TO THE RUSSIAN CAPITAL OF ST. PETERSBURG. THE CITY'S MAIN STREET, NEVSKY PROSPEKT, IS SHOWN HERE AROUND 1900. BECAUSE OF HIS OBVIOUS INTELLIGENCE, MENDELEYEV WAS ACCEPTED INTO HIGH SCHOOL ON A FULL SCHOLARSHIP.

her children. Upon arriving in Moscow, her hopes were dashed as Dmitry was not accepted into any schools. He was not turned down for lack of intelligence, but rather because of old rules in Moscow's educational system. Children from Siberia were considered lower class and were not accepted as students. Mariya was heartbroken but determined. She moved the family another 400 miles

(640 km) to St. Petersburg, the capital of Russia, where Dmitry was accepted into school on a full scholarship.

ILLNESS STRIKES THE MENDELEYEVS

Within a few months of their arrival in St. Petersburg, Mariya Mendeleyev contracted tuberculosis, which is a

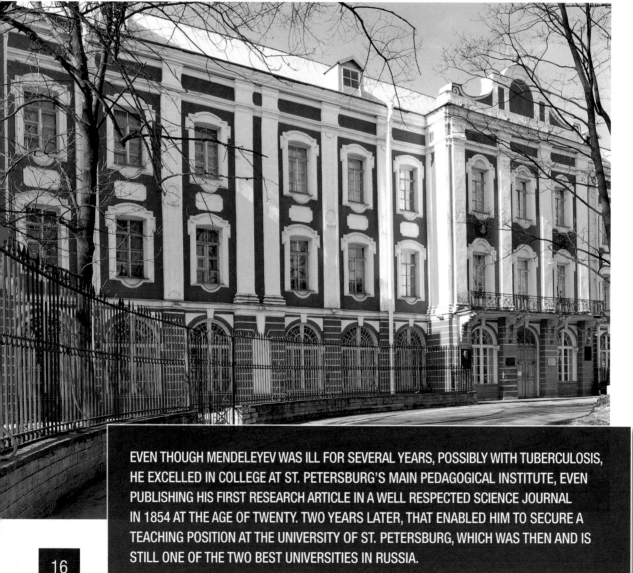

EVEN THOUGH MENDELEYEV WAS ILL FOR SEVERAL YEARS, POSSIBLY WITH TUBERCULOSIS, HE EXCELLED IN COLLEGE AT ST. PETERSBURG'S MAIN PEDAGOGICAL INSTITUTE, EVEN PUBLISHING HIS FIRST RESEARCH ARTICLE IN A WELL RESPECTED SCIENCE JOURNAL IN 1854 AT THE AGE OF TWENTY. TWO YEARS LATER, THAT ENABLED HIM TO SECURE A TEACHING POSITION AT THE UNIVERSITY OF ST. PETERSBURG, WHICH WAS THEN AND IS STILL ONE OF THE TWO BEST UNIVERSITIES IN RUSSIA.

very serious illness that was common in the mid-1800s, and she died. A year later, Dmitry's sister Liza also died from the disease. And in 1853, Dmitry received the same diagnosis. Doctors told him he had only a few months to live. It was Dmitry's third year at the Main Pedagogical Institute, and already he was starting to build a good name. He was known as a devoted science student, and many of his professors thought he was truly gifted.

Even though he was sick, Dmitry still pursued his studies. If anything, he was even more devoted. Even though he was forced to spend a lot of time in bed, whenever Dmitry wasn't resting, he was studying science. At the time, Dmitry was moving away from his initial interest in physics and was beginning to concentrate much more on chemistry.

Mendeleyev was far from alone in his new focus. During the mid-1800s, scientists were beginning to discover many new elements. Each new element provided another piece to the puzzle of the matter that makes up the universe. To help him stay in school, Dmitry's friends brought his assignments to his bedside, where he worked diligently. Eventually, Dmitry even ventured out and began going to the school's laboratory to work on his own experiments. He would return to his bed, exhausted, and write up his experiments before falling asleep. By the time he was twenty, Dmitry was publishing articles on his own experiments. His first article, published in a well-respected science journal in 1854, was called "Chemical Analysis of a Sample from Finland."

MEMORABLE LAST WORDS

According to *Mendeleyev's Dream: The Quest for the Elements*, written by Paul Strathern, Mariya Mendeleyev's dying words to her son Dmitry were these: "Refrain from illusions, insist on work and not on words. Patiently seek divine and scientific truth." Thirty years later, Mendeleyev quoted his mother's words in a scientific paper, which he dedicated to her.

Publishing a journal article was a huge accomplishment for such a young scientist!

A LIFE-CHANGING EVENT

In 1855, the Main Pedagogical Institute recognized Dmitry with the Student of the Year Award. This was a great and well-deserved honor. Along with the award, Dmitry also received a teaching position. His first teaching post was at a school in Simferopol in the Crimean Peninsula. The move turned out to be a life-changing event, but not in the way Dmitry expected.

The Crimean Peninsula lies to the south of Russia and juts into the Black Sea. Unfortunately, when Mendeleyev arrived there, the Crimean War had closed the school for months. So Dmitry found himself poor and living in a war-torn country.

WHEN MENDELEYEV ARRIVED IN SIMFEROPOL ON THE CRIMEAN PENINSULA TO TAKE HIS FIRST TEACHING POSITION, RUSSIA WAS AT WAR THERE AGAINST AN ALLIANCE OF THE BRITISH EMPIRE, FRANCE, AND THE OTTOMAN EMPIRE. THE SCHOOL WAS CLOSED, BUT THE TRIP HAD A SIDE BENEFIT. A DOCTOR THERE DIAGNOSED HIS ILLNESS AS NONFATAL. HE RETURNED TO ST. PETERSBURG WITH A NEW LEASE ON LIFE.

Dmitry thought his journey was turning out to be worthless, but about one month after his arrival, a doctor there diagnosed his disease as nonfatal. Instead of suffering from a life-threatening illness, he was on the road to recovery. He left the peninsula and headed back to St. Petersburg, eager to begin a career. He hoped to make great contributions to chemistry, but not even he could foresee that his work would launch a scientific revolution.

CHAPTER TWO

THE MAKING OF A CHEMIST

Arriving in St. Petersburg with a Student of the Year Award from the Main Pedagogical Institute, Mendeleyev was able to get a teaching job at the prestigious University of St. Petersburg (today's St. Petersburg State University). But building a university career is never an easy task. Would the twenty-two-year-old professor have the tools for success? The answer would turn out to be a definite yes!

Becoming a professor at a distinguished university was a remarkable accomplishment for such a young man. It also gave him time to pursue the research questions that excited him. A few days each week he would give lectures on chemistry, with the remainder of the week open so he could perform experiments using the school's laboratories.

A NEW SCIENCE

The mid-nineteenth century was an exciting time for chemistry. In many ways, chemistry was a new science because it had been fewer than fifty years since the 1808 publication of *A New System of Chemical Philosophy* by John Dalton (1766–1844).

As the nineteenth century began, Dalton, who had started out studying meteorology, had turned his interests to the properties of gases. One particularly interesting property was that when two gases reacted together to form a new gas, the volumes of the combining gases always were in a simple ratio. For example, two liters of hydrogen and one liter of oxygen combined to form two liters of water vapor. This property is now known as Gay-Lussac's law.

By 1803, Dalton realized that property fit with an idea

JOHN DALTON REVOLUTIONIZED CHEMISTRY IN 1803 WITH HIS ATOMIC THEORY AND THE IDEA OF CHEMICAL ELEMENTS. BY MENDELEYEV'S TIME, MANY NEW ELEMENTS HAD BEEN DISCOVERED AND WERE CONTINUING TO BE DISCOVERED AT A RAPID RATE. HIS PERIODIC TABLE BROUGHT ABOUT A NEW REVOLUTION BY NOT ONLY EXPANDING ON DALTON'S ORIGINAL TABLE OF THE ELEMENTS, SHOWN HERE, BUT ALSO PROVIDING A SCHEME FOR ORGANIZING THE ELEMENTS—DISCOVERED AND UNDISCOVERED—ACCORDING TO THEIR PROPERTIES.

of the ancient Greek philosophers Leucippus and Democritus. Twenty-three centuries earlier, they imagined cutting up a piece of matter until it was *atomos*, meaning "indivisible." Democritus wrote that there is a limit to how small a piece of matter can be cut and still remain the same substance. That turned out to be right, but most substances are compounds. Thus the smallest possible piece of most substances is usually a molecule instead of an atom.

However, molecules can be divided into their atoms, and that is where chemistry enters the picture. Dalton was the first to bring the notion of atoms into interpretation of laboratory observations.

As you might imagine, the road from the philosophy of the ancient Greeks to the chemistry of Dalton was a long one, with many interesting twists and stops along the way. It began with an activity called alchemy, in which people tried to make certain substances out of others, often by heating things together. Most often, alchemists were searching for ways to turn less valuable metals into gold. Scientists now know that their techniques were doomed to failure. Gold is an element, and neither ancient alchemy nor modern chemistry can change one kind of atom into another.

Though many alchemists were frauds, others succeeded in developing a rudimentary knowledge of matter, extracting or purifying many useful elements and compounds from natural substances. By the seventeenth century, scientific thinking had begun to take hold, and alchemy gradually transformed into the science of chemistry. Eighteenth-century chemists made a number of important discoveries, including facts about the behavior of gases, the processes of combustion and corrosion, and the relationship between electricity and matter. None of those phenomena were fully understood, but plenty of evidence and measurements were being gathered systematically and scientifically.

A New System of Chemical Philosophy was a breakthrough on the theory of atoms. Dalton wrote that every substance was composed of tiny bits of matter called atoms. If a substance's atoms were all the same, it was an element. Other substances called compounds were made of molecules, or combinations of atoms in specific ratios, such as water (two hydrogen atoms combined with one oxygen atom) and table salt (one sodium atom plus one chlorine atom to form sodium chloride).

The book also stated that atoms of different elements have different properties, including their weight; and when atoms join to form compounds, it is always in small whole numbers of each atom—no fractional atoms allowed. That idea explained Gay-Lussac's law and other properties of gases, as well as many other important discoveries of the previous century.

Using those simple rules, Dalton was able to determine the atomic weight of different elements. He assigned hydrogen, the lightest element, one unit of atomic weight and determined the atomic weight of other atoms from that. For instance, he knew that water was a compound of hydrogen and oxygen, with eight times as much oxygen by weight. Assuming that a water molecule had one atom of each, he set the atomic weight of oxygen to eight units. Later on, when more research showed that water molecules had two atoms of hydrogen and one of oxygen, scientists corrected that result, setting the atomic weight of oxygen to sixteen.

Following Dalton's method, scientists studying chemical reactions gradually identified more compounds and the elements that composed them, and they determined the atomic weights of each element. Though no one had detected individual atoms, Dalton's idea of elements and compounds, atoms and molecules, had given chemistry a new basic vocabulary.

MENDELEYEV'S QUESTIONS

As is often the case with scientific breakthroughs, atomic theory also opened up a wealth of new questions. The young Professor Mendeleyev selected these for his own research: How many elements are there, what are their properties, and is there a way to classify them according to their chemical and physical properties?

In the mid-nineteenth century, new elements were being discovered almost every year. Among the most recently discovered when Mendeleyev began his work

WHY RUSSIA USED A DIFFERENT CALENDAR

During Mendeleyev's time, Russia was one of the last countries in the world still using the ancient Julian calendar, named for Julius Caesar. It was almost the same as the current calendar except that it had leap years every four years without fail. The rest of the world was using the Gregorian calendar, which was introduced in 1582 by Pope Gregory XIII and is the one people still use today. In century years, like 1700, 1800, and 1900, the Gregorian calendar does not have leap years. Only when the century year is divisible by 400, such as 1600 and 2000, does the Gregorian calendar add February 29.

The reason for the change was that the calendar no longer matched the seasons correctly. Not every country introduced the Gregorian calendar at the same time. Countries that had once been ruled by Roman Catholic monarchs but were now led by Protestant or Orthodox Christian rulers refused to follow orders from the pope. Still, the new calendar made scientific sense. It was adopted in the British Empire, including the American colonies, in 1752. Russia did not adopt the Gregorian calendar until the 1917 Russian Revolution deposed the Orthodox Christian monarchy.

at the University of St. Petersburg were aluminum and bromine. Like sodium and chlorine, those elements are both quite common on Earth, but they occur naturally only in compounds. Mendeleyev eagerly read scientific journals to follow the research of the scientists who were discovering new elements and determining their properties. Chemists from all over the world were making discoveries, and he was determined to join them.

For around six months in 1858, Mendeleyev taught at the University of St. Petersburg while working in the school's laboratories. But, as his knowledge of chemistry grew, he became frustrated with Russia's place in the scientific world. Mendeleyev enjoyed teaching but had a hard time finding many Russian scientists who were performing more daring experiments that challenged the existing knowledge. Eventually, Mendeleyev began to see that Russia was lagging behind the rest of the world in science.

Russia was even using an old calendar that was twelve days behind the rest of the world! For example, when it was July 25, 1868, in France, it was only July 13, 1868, in Russia. More important, Russia was behind the world politically as well. The country was still using an old political system called feudalism, which places land and power in the hands of very few people and leaves most of the population very poor and unable to work toward a better life.

In terms of science, Russia had very little to offer Mendeleyev. While the rest of the world was booming with opportunities in chemistry, Russia did not have as many science-related jobs or science-oriented universities. Frustrated, Mendeleyev finally applied to study abroad for two years.

NEW OPPORTUNITIES

In 1859, Mendeleyev received an opportunity to study with the renowned chemist Henri-Victor Regnault (1810–1878) in Paris, France. Regnault was best known for his careful measurements in thermodynamics, the study of heat flow and temperature, especially the heat- and temperature-related properties of matter.

After studying with Regnault for a few months, Mendeleyev headed off to Heidelberg, Germany, to study with Robert Bunsen (1811–1899) and Gustav Kirchhoff (1824–1887). These two brilliant scientists were partners and shared a state-of-the art laboratory. When Mendeleyev joined them, they were developing ways to use a device called

FRUSTRATED WITH THE STATE OF SCIENTIFIC RESEARCH IN RUSSIA, MENDELEYEV LOOKED FOR OPPORTUNITIES ABROAD. HE FIRST STUDIED WITH RENOWNED FRENCH CHEMIST HENRI-VICTOR REGNAULT, SHOWN HERE, BEFORE MOVING TO HEIDELBERG, GERMANY, TO STUDY SPECTROSCOPY (USING THE COLORS OF LIGHT TO ANALYZE A SUBSTANCE) WITH ROBERT BUNSEN AND GUSTAV KIRCHHOFF.

the spectroscope to analyze the chemical makeup of a substance.

In the first few weeks of his visit, Mendeleyev spent long hours studying with Kirchhoff and Bunsen at their laboratories. Along with the spectroscope, Bunsen was working on a special burner to use during experiments. Now, people know this invention as the Bunsen burner, and it is still one of the most common tools in laboratories to this day. Mendeleyev felt motivated just being around these two brilliant minds.

SPECTROSCOPY: SHEDDING LIGHT ON CHEMISTRY

One technique that nineteenth-century chemists used to identify elements was the flame test. When a substance is heated in a flame, like the one produced by a Bunsen burner, it glows with a characteristic color. Because the human eye can be fooled by similar colors, Bunsen and Kirchhoff used a spectroscope to analyze the glow.

A spectroscope uses a prism or another device to spread light out into a spectrum, or the colors that make it up. When sunlight passes through a spectroscope, it produces a continuous band of colors, like a rainbow. But when the light from a glowing hot substance passes through, it produces a series of sharp lines instead. Each element produces its own distinct set of lines. One day in 1861, when Bunsen and Kirchhoff heated and analyzed a new chemical sample, they saw a set of lines that they had never seen before. They suspected that they had found a new element. Other chemical tests confirmed their suspicion, and they named the new element rubidium.

Unfortunately, Mendeleyev was also very moody and stubborn.

As the weeks passed, Mendeleyev and Bunsen started to fight. At first, the pair fought only once in a while, but soon they were fighting every day. One day, in the middle of an argument, Mendeleyev stormed out of Bunsen and Kirchhoff's laboratory and vowed never to return.

MENDELEYEV ON HIS OWN

Mendeleyev set up his own laboratory in Heidelberg. It was nothing more than a small room jammed with lab equipment. In some ways, Mendeleyev's decision to walk out of Bunsen's laboratory was hard to understand. He was still quite young. He had learned a lot from these two scientists and still had so much more to learn. But his bitterness toward his former mentor drove him to work exceptionally hard in his new home laboratory. And he soon began to produce results.

At first, Mendeleyev performed simple experiments on solubility—the amount of a substance that will dissolve in a given amount of another substance. He began working with alcohol and water, but he soon was digging deeper and studying the way chemicals in solution reacted with each other. Mendeleyev's experiments were determining the valence of different elements. The valence of an atom determines the way it combines with other atoms to form new substances. Some atoms have a positive valence and other have a negative valence. They combine to form a compound with valence zero.

For instance, an atom of sodium (valence +1) combines with one atom of chlorine (valence -1) to form sodium chloride (valence +1-1=0). But an atom

of calcium (valence +2) combines with two atoms of chlorine to make calcium chloride (valence +2-1-1 =0). Soon Mendeleyev was publishing papers on his experiments with elements and valence; he suggested that figuring out an element's valence was just as important as figuring out its other physical and chemical properties. Many scientists agreed and turned their attention to valence.

Mendeleyev's mind was brilliant when it came to chemistry. He could not only

MENDELEYEV HAD STRONG IDEAS, AND HIS PERSONALITY CLASHED WITH BUNSEN'S. THOUGH IT SEEMED SHORTSIGHTED FOR THE YOUNG SCIENTIST TO LEAVE WHEN HE STILL HAD SO MUCH TO LEARN, MENDELEYEV DECIDED TO SET UP HIS OWN LABORATORY. THE MOVE FREED HIS BRILLIANT MIND TO INVESTIGATE NEW ELEMENTS AND TO DISCOVER PATTERNS IN THEIR PROPERTIES. HE WAS SOON PUBLISHING HIS RESULTS IN IMPORTANT JOURNALS AND BUILDING A SUBSTANTIAL REPUTATION.

remember a vast amount of complicated scientific information, but he could apply this information to come up with new ideas and experiments. Mendeleyev's mind worked in a constant process of trial and error. Each experiment he performed led him further. Many times, Mendeleyev's experiments were based on initial hunches, but these hunches came from his ability to think through complex scientific information.

Especially valuable was his natural ability to recognize and discover patterns. Each experiment with a new solution led to new discoveries about the elements in them. The more experiments he performed, the more new patterns began to emerge in his mind. Mendeleyev reported his results in scientific journals. But he never lost track of the bigger picture: What are the relationships among all the patterns? Is there a larger arrangement to classify the elements? How do they all fit together?

CHAPTER THREE

PROFESSOR MENDELEYEV

*I*n 1861, Mendeleyev returned to St. Petersburg, where he married Feozva Leshcheva. Then, in 1864, he accepted a teaching position at the Technological Institute in St. Petersburg and began to study for his Ph.D., or doctoral degree, at the University of St. Petersburg. To earn a Ph.D., a student must complete an original research project. His work in Germany showed that he could certainly do that, but he needed to complete the degree to qualify as a university professor.

By 1865, had earned his Ph.D. and began teaching at the university. Remarkably, only a year later, he was named chair of the university's chemistry department. He and his wife had their first child in 1866, a son they named Volodya. Soon after, they had a daughter named Olga.

IN 1865, MENDELEYEV COMPLETED HIS PH.D. DEGREE. THAT EARNED HIM THE RIGHT TO WEAR THE DOCTORAL CAP, GOWN, AND HOOD FOR ACADEMIC CEREMONIES, SHOWN IN THIS PICTURE. THE DEGREE QUALIFIED MENDELEYEV FOR A POSITION AS A PROFESSOR AT THE UNIVERSITY OF ST. PETERSBURG, WHERE A YEAR LATER, HE WAS NAMED CHAIR OF THE CHEMISTRY DEPARTMENT.

THE SEARCH FOR A PATTERN

By 1869, Mendeleyev had been teaching at the University of St. Petersburg for six years. He had come into his own as a professor and furthered his reputation as a brilliant scientist. Mendeleyev enjoyed teaching and it showed. His students loved him. They said he did not merely teach chemistry; he made it fun.

He clearly was also enjoying his research. His reputation as a scientist grew as he published numerous scientific articles on his findings. He also wrote a textbook called *The Principles of Chemistry.* After it was published, Mendeleyev saw the need for a second volume. His work on that

A LIBRARY OF
UNIVERSAL LITERATURE

IN FOUR PARTS

Comprising Science, Biography, Fiction
and the Great Orations

PART ONE—SCIENCE

The Principles of Chemistry

(PART ONE)

BY
D. MENDELÉEFF

NEW YORK
P. F. COLLIER AND SON
·MCMI·
25

MENDELEYEV'S FIRST PART OF *THE PRINCIPLES OF CHEMISTRY* WAS THE MAJOR STEPPING-STONE ON HIS PATH TO DISCOVERING THE PERIODIC TABLE OF ELEMENTS. AFTER TEACHING AT THE UNIVERSITY OF ST. PETERSBURG FOR SEVERAL YEARS, MENDELEYEV REALIZED HE DID NOT POSSESS A TEXTBOOK THAT COULD TEACH HIS STUDENTS TO PROPERLY UNDERSTAND THE RELATIONSHIPS BETWEEN THE ELEMENTS. IN THE MID-1860S, MENDELEYEV BEGAN WRITING A TEXT THAT WOULD MEET THESE NEEDS. THE RESULT WAS *THE PRINCIPLES OF CHEMISTRY*, WHICH APPEARED IN FOUR PARTS. THE FIRST PART APPEARED IN 1868, THE SAME YEAR HE BECAME A FOUNDING MEMBER OF THE RUSSIAN CHEMICAL SOCIETY. THIS 1901 EDITION WAS ONE OF THE FIRST PRINTINGS TO INCLUDE ALL FOUR PARTS OF MENDELEYEV'S GROUNDBREAKING WORK.

volume would lead him to a breakthrough idea that would revolutionize chemistry.

The Principles of Chemistry was his passion. Mendeleyev put every ounce of his knowledge into this book. As Paul Strathern explains in *Mendeleyev's Dream: The Quest for the Elements*, Mendeleyev's text became one of the most important and detailed books published on chemistry during the time. *Principles* was so carefully researched that its reference section ended up being as long as the book itself. It remained the most popular and definitive book on chemistry for more than thirty years!

By February 1869, Mendeleyev had already written the first volume of the book and he was now working on volume two. Mendeleyev finished the first two chapters on the alkali metal group (the elements that include sodium and potassium). However, chapter three posed a problem because before he could write it, he had to figure out which elements had properties closest to the alkali metals. He was stuck.

When Mendeleyev sat down to write that chapter, he felt the pressure of his approaching self-imposed deadline. Mendeleyev had allowed himself one last weekend to meet the deadline to complete chapter three. Unfortunately, he was going out of a town for a few days to give a lecture and to tour the Russian countryside. He knew he had to make a lot of progress on his book over the weekend, but he was feeling a little overwhelmed about how to continue.

One of the problems facing Mendeleyev was that he had noticed a big problem with the elements in the entire field of chemistry. By 1869, chemists had already discovered 63 elements. (As of 2013, 118 elements have been found in nature or made artificially.) Although scientists recognized many similarities in properties among various groupings, no one had found a way to organize all the elements on a chart that demonstrated the relationships among them. Mendeleyev could not write chapter 3 of his book without first organizing the elements.

Mendeleyev's deadline came and went, and his frustration grew. He paced through his office, making notes, muttering, and thinking as hard as he could. After a few

IN 1869, MENDELEYEV FINALLY HAD THE INSIGHTS HE NEEDED TO ARRANGE ALL THE KNOWN ELEMENTS IN A PERIODIC TABLE. THIS IS HIS HANDWRITTEN DRAFT, FULL OF NOTATIONS AND CORRECTIONS, OF THE ARRANGEMENT THAT WAS EVENTUALLY PUBLISHED. NOTE THE DATE, FEBRUARY (II) 17, 1869, IN THE LOWER LEFT.

SOME ELEMENTS DISCOVERED BY 1869

Aluminum (Al): Discovered in 1825 by Hans Christian Oersted in Denmark

Boron (B): Discovered in 1808 by J. L. Gay-Lussac and L. J. Thenard as well as Sir Humphry Davy

Calcium (Ca): Discovered in 1808 by Sir Humphry Davy in London, England

Carbon (C): Discovered in prehistoric time

Hydrogen (H): Discovered in 1766 by Henry Cavendish in London, England

Iron (Fe): Discovered by ancient civilizations

Lithium (Li): Discovered in 1817 by J. A. Arfwedson in Sweden

Nitrogen (N): Discovered in 1772 by Daniel Rutherford in Edinburgh, Scotland, as well as in the early 1770s by Carl Wilhelm Scheele in Sweden, Henry Cavendish, and Joseph Priestly in England

Oxygen (O): Discovered independently around 1772 by Carl Wilhelm Scheele in Sweden and 1774 by Joseph Priestley in England

Zinc (Zn): Known in India and China before 1500 and to the Greeks and Romans before 20 BCE

hours, his office was a mess. There were papers everywhere. He had notes scribbled in the margins of books, on random pieces of paper, even in old notebooks he had used while a student at the University of St. Petersburg. He was determined to figure out how to organize all of the elements, but it continued to elude him.

He knew there had to be a pattern that tied all of the elements together. He found a paper written in 1862 on this topic by French geologist Alexandre-

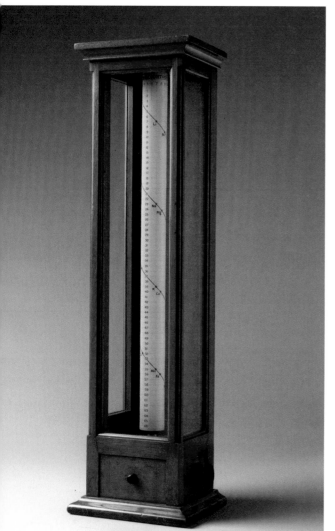

Émile-Beguyer de Chancourtois, who thought he had figured out part of the pattern. De Chancourtois had created a list of the elements arranged by increasing atomic weight, and he noted a repeating pattern in some of their physical properties, such as melting point, boiling point, density (their weight per cubic centimeter or cubic inch), and the kind of minerals they formed.

De Chancourtois devised a spiral chart of the elements that when wrapped on a cylinder had an interesting feature. Similar elements were in a vertical line down the cylinder. He presented

MENDELEYEV'S PERIODIC TABLE WAS INSPIRED BY THIS 1862 CHART BY FRENCH GEOLOGIST ALEXANDRE-ÉMILE-BÉGUYER DE CHANCOURTOIS, WHO CLASSIFIED THE ELEMENTS BY ATOMIC WEIGHTS AND ORDERED THEM ON A SPIRAL AROUND A CYLINDER DIVIDED INTO SIXTEEN PARTS. TELLURIUM WAS AT THE CENTER OF HIS SYSTEM, SO DE CHANCOURTOIS NAMED IT THE TELLURIC SCREW. MENDELEYEV THOUGHT THIS CHART WAS ON THE RIGHT TRACK, BUT IT DID NOT GROUP THE ELEMENTS CORRECTLY ACCORDING TO THEIR CHEMICAL PROPERTIES.

his findings to the French Academy of Sciences. Unfortunately, the published version of his presentation left out his diagram. Mendeleyev thought de Chancourtois was on the right track, but de Chancourtois's arrangement did not group elements correctly according to their chemical properties.

English chemist John A. R. Newlands had also researched how to group the elements. In a paper published in 1864, Newlands noted that of the sixty known elements, chemical groups (such as alkali metals) repeated every eight elements. He named this the law of octaves and compared it to a musical scale. With the work of both of these scientists in mind, Mendeleyev began plotting out all of the elements on a piece of paper. When his first version didn't make sense, he made a new version. Then, he plotted a newer version and so on. Over the next two days, Mendeleyev designed more than ten different ways to organize the known elements.

None of them worked, but he was determined not to give up until he discovered the right arrangement.

CHAPTER FOUR

SUCCESS!

Mendeleyev refused to give up. Every time he tried an arrangement of the elements and failed, he realized he was still missing some pieces of the puzzle. But he also realized he was a step closer to finding a pattern that worked. He was a man so obsessed that he worked day and night without a break.

Finally on the morning of February 17, 1869, Mendeleyev took a break for breakfast. He was supposed to leave on a train right after breakfast to travel to the town of Tver, visit the Russian countryside, and deliver a lecture to the workers of a cheese-making factory. After eating, he returned to his study and opened the letter of invitation from the Voluntary Economic Cooperative of Tver that detailed the trip. This letter would soon become a piece of scientific history.

GROUPING BY CHEMICAL PROPERTIES

Mendeleyev began jotting some notes on the back of the letter. He wrote the chemical symbols and atomic weight for four elements that had similar properties: fluorine (F), chloride (Cl), bromine (Br), and iodine (I). These four elements all reacted with metals to form salts. Chemists named them halogens, meaning "salt formers." Mendeleyev also noted that they had a valence of +1. Mendeleyev began to see a pattern—the elements could be grouped not only by atomic weight but also by a chemical property!

Mendeleyev then did the same thing for another group of elements with similar chemical properties he called the oxygen elements: oxygen (O), sulfur (S), selenium (Se), and tellurium (Te). His computations looked like this:

Chemical Symbol = Atomic Weight

F = 19	Cl = 35	Br = 80	I = 127
O = 16	S = 32	Se = 79	Te = 128
N = 14	P = 41	As = 75	Sb = 122

This chart shows a pattern when the elements are grouped by their atomic weight and by their similar properties. Mendeleyev noticed that almost all of the

CHEMICAL SYMBOLS

Whenever a new element is discovered, scientists assign that element a symbol, or abbreviation. A scientific symbol is really just a shorter way of representing a longer, more complex word. For example, the symbol for the element phosphorus is P. Every element has a symbol. Oxygen is denoted by the letter O, sulfur by S, carbon by C, and so forth. Some elements have symbols from their names in other languages. Ag (*argentum*) for silver, Au (*aurum*) for gold, Cu (*cuprum*) for copper, Fe (*ferrum*) for iron, Pb (*plumbum*) for lead, and Sn (*stannum*) for tin all come from Latin. Tungsten gets its name from the Swedish words *tung sten*, meaning heavy stone, but its symbol W (*wolfram*) comes from the name of a mineral that contains it, wolframite (from the German words *wolf rahm*, meaning "wolf soot").

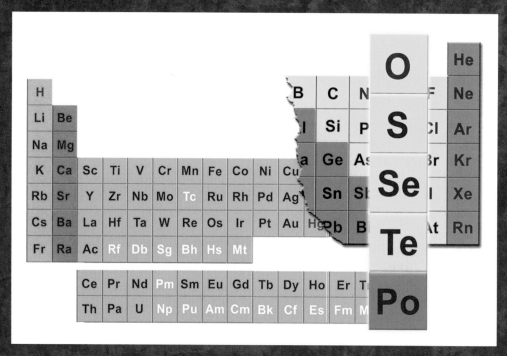

atomic weights get smaller going down the column and the atomic weights get larger going across the rows. A pattern was beginning to form in his mind.

A FLASH OF INSPIRATION

When Mendeleyev's friend A. A. Inostrantzev stopped by on the morning of February 17, he found the scientist working intently, but extremely frustrated and upset as well. Mendeleyev told Inostrantzev that he had been working for three days without even breaking for sleep. He knew he was close to making a great discovery but he just could not figure it out. How did the elements fit together?

By this time, Mendeleyev had missed his train to the countryside, but it did not matter to him. He knew that he was close and could not stop now. As February 17 came to a close, Mendeleyev still had not come up with a way to organize the elements. His mind and body were overcome with fatigue, but he continued to work in his study, preoccupied with finding the solution. As the night passed, Mendeleyev's body gave up on him. He fell asleep in his study, his head resting in his arms atop his desk.

Even in his sleep, Mendeleyev's brain kept working on the problem. And when he awoke, according to popular accounts, he had the answer: When the elements were listed in order of their atomic weights, their chemical and physical properties repeated in a

WHAT HAPPENS WHEN PEOPLE DREAM?

The story of Mendeleyev's dream is not surprising to scientists who study sleep. Research shows that while a person is dreaming, he or she processes an enormous amount of information. Dreams are a way for a person to process daily experiences. Even while asleep, the brain has periods called REM sleep (for rapid eye movement) when it actively processes thoughts and emotions, including any information taken in during the day. Its intense activity during REM sleep ranges from decision making to learning how to do new things, like playing a musical instrument. In Mendeleyev's case, his dreams may have shown him the way to a historic discovery!

series. Because it seemed to come to him in a dream, Mendeleyev's idea seemed to be a flash of inspiration. But like so many flashes of inspiration, it could only have come to a mind that was prepared by hard work to see it.

FILLING IN THE GAPS

According to the Russiapedia Web site, the story of Mendeleyev's dream is probably not true. But there is little doubt that discovering the periodic table required both insight and hard work. Part of Mendeleyev's insight came from realizing that there were surely more elements to be discovered. It had been less than ten years since he had worked with Bunsen and Kirchhoff, and he had seen the number of elements grow from the low sixties to sixty-nine. So when he arranged each group of similar elements in a row by increasing atomic weight, he realized he had to leave some gaps so that the elements also lined up in columns with increasing atomic weights as well.

After two weeks of arranging and rearranging elements, Mendeleyev was satisfied with the chart he had created. He published the periodic table of

Tabelle I.

Typische Elemente

							der chemischen Elemente.
			K = 39	Rb = 85	Cs = 133	—	—
			Ca = 40	Sr = 87	Ba = 137	—	—
			—	?Yt = 88?	?Di = 138?	Er = 178?	—
			Ti = 48?	Zr = 90	Ce = 140?	?La = 180?	Th = 231
			V = 51	Nb = 94	—	Ta = 182	—
			Cr = 52	Mo = 96	—	W = 184	U = 240
			Mn = 55	—	—	—	—
			Fe = 56	Ru = 104	—	Os = 195?	—
			Co = 59	Rh = 104	—	Ir = 197	—
			Ni = 59	Pd = 106	—	Pt = 198?	—
H = 1	Li = 7	Na = 23	Cu = 63	Ag = 108	—	Au = 199?	—
	Be = 9,4	Mg = 24	Zn = 65	Cd = 112	—	Hg = 200	—
	B = 11	Al = 27,3	—	In = 113	—	Tl = 204	—
	C = 12	Si = 28	—	Sn = 118	—	Pb = 207	—
	N = 14	P = 31	As = 75	Sb = 122	—	Bi = 208	—
	O = 16	S = 32	Se = 78	Te = 125?	—	—	—
	F = 19	Cl = 35,5	Br = 80	J = 127	—	—	—

THIS IS AN EARLY VERSION OF MENDELEYEV'S PERIODIC TABLE OF THE ELEMENTS WITH SOME ADJUSTMENTS FROM HIS ORIGINAL VERSION IN WHICH A FEW OF THE ATOMIC WEIGHTS SEEMED TO BE OUT OF ORDER. MODERN PERIODIC TABLES REVERSE THE ROWS AND COLUMNS AND HAVE THE GAPS FILLED IN BY ELEMENTS THAT WERE DISCOVERED SINCE THEN. IT CONTAINS EVERY GROUP OF ELEMENTS KNOWN TODAY EXCEPT FOR THE NOBLE GASES, WHICH WERE UNKNOWN AT THAT TIME.

elements in a paper called "An Attempt at a System of the Elements Based on Their Atomic Weight and Chemical Affinity." Going from the top to bottom of each column and following columns from left to right, the atomic weights increased, while similar elements, such as the halogens and the oxygen group, appeared in rows. In other words, following the chart by increasing atomic number will reproduce a periodically repeating sequence of physical and chemical properties.

In a few places, the atomic weights seemed to be out of order (for example Au or gold 197 and Bi or bismuth 210 were between Hg or mercury 200 and Tl or thallium 204), but scientists weren't always sure of their measurements of atomic weights or of chemical groups at that time. So the first periodic table had gaps and question marks, but chemists recognized it as a breakthrough in classifying the growing number of chemical elements.

The next year, Mendeleyev took his table even further. As the American Institute of Physics' Center for the History of Physics Web site states, "Mendeleev was bold enough to suggest that new elements not yet discovered would be found to fill the blank places. He even went so far as to predict the properties of the missing elements." Many scientists were skeptical at first, but then some missing elements were discovered with properties that matched Mendeleyev's predictions: gallium in 1875, scandium in 1879, and

Periodic Table

Period ↓ / Group →	1	2	3	4	5	6	7	8	9	10	11	12	13	14	15	16	17	18
1	1 **H** 1.008																	2 **He** 4.003
2	3 **Li** 6.941	4 **Be** 9.012											5 **B** 10.81	6 **C** 12.01	7 **N** 14.01	8 **O** 16	9 **F** 19	10 **Ne** 20.18
3	11 **Na** 22.99	12 **Mg** 24.31											13 **Al** 26.98	14 **Si** 28.09	15 **P** 30.97	16 **S** 32.07	17 **Cl** 35.45	18 **Ar** 39.95
4	19 **K** 39.10	20 **Ca** 40.08	21 **Sc** 44.96	22 **Ti** 47.88	23 **V** 50.94	24 **Cr** 52	25 **Mn** 54.94	26 **Fe** 55.85	27 **Co** 58.47	28 **Ni** 58.69	29 **Cu** 63.55	30 **Zn** 65.39	31 **Ga** 69.72	32 **Ge** 72.59	33 **As** 74.92	34 **Se** 78.96	35 **Br** 79.9	36 **Kr** 83.8
5	37 **Rb** 85.47	38 **Sr** 87.62	39 **Y** 88.91	40 **Zr** 91.22	41 **Nb** 92.91	42 **Mo** 95.94	43 **Tc** (98)	44 **Ru** 101.1	45 **Rh** 102.9	46 **Pd** 106.4	47 **Ag** 107.9	48 **Cd** 112.4	49 **In** 114.8	50 **Sn** 118.7	51 **Sb** 121.8	52 **Te** 127.6	53 **I** 126.9	54 **Xe** 131.3
6	55 **Cs** 132.9	56 **Ba** 137.3	57 **La** 138.9	72 **Hf** 178.5	73 **Ta** 180.9	74 **W** 183.9	75 **Re** 186.2	76 **Os** 190.2	77 **Ir** 192.2	78 **Pt** 195.1	79 **Au** 197	80 **Hg** 200.5	81 **Tl** 204.4	82 **Pb** 207.2	83 **Bi** 209	84 **Po** (210)	85 **At** (210)	86 **Rn** (222)
7	87 **Fr** (223)	88 **Ra** (226)	89 **Ac** (227)	104 **Rf** (257)	105 **Db** (260)	106 **Sg** (263)	107 **Bh** (262)	108 **Hs** (265)	109 **Mt** (266)	110 **Ds** (271)	111 **Rq** (272)	112 **Uub** (285)	113 **Uut** (284)	114 **Fl** (289)	115 **Uup** (288)	116 **Lv** (292)	117 **Uus** 0	118 **Uuo** 0

Legend: Nonmetals · Alkali metals · Alkaline Earth metals · Transition elements · Other metals · Metalloids · Halogenes · Noble gases · Lanthanides · Actinides

6	58 **Ce** 140.1	59 **Pr** 140.9	60 **Nd** 144.2	61 **Pm** (147)	62 **Sm** 150.4	63 **Eu** 152	64 **Gd** 157.3	65 **Tb** 158.9	66 **Dy** 162.5	67 **Ho** 164.9	68 **Er** 167.3	69 **Tm** 168.9	70 **Yb** 173	71 **Lu** 175
7	90 **Th** 232	91 **Pa** (231)	92 **U** (238)	93 **Np** (237)	94 **Pu** (242)	95 **Am** (243)	96 **Cm** (247)	97 **Bk** (247)	98 **Cf** (249)	99 **Es** (254)	100 **Fm** (253)	101 **Md** (256)	102 **No** (254)	103 **Lr** (257)

THIS MODERN PERIODIC TABLE HAS ALL THE MISSING GAPS FROM MENDELEYEV'S TIME FILLED IN. EACH SQUARE HAS THE ATOMIC SYMBOL IN THE CENTER, THE ATOMIC NUMBER IN THE UPPER LEFT, AND THE ATOMIC WEIGHT AT THE BOTTOM. IT ALSO INCLUDES SEVERAL ADDITIONS THAT MENDELEYEV DID NOT INCLUDE IN HIS ORIGINAL TABLE, INCLUDING THE NOBLE GASES (COLUMN 18) AND ARTIFICIALLY CREATED ELEMENTS SUCH AS ATOMIC NUMBER 101, MENDELEYEVIUM (MD). NOTE THAT ROWS 6 AND 7 HAVE EXTRA ELEMENTS ADDED AT THE BOTTOM. THEY ALL BELONG IN COLUMN THREE WITH LANTHANUM (LA) OR ACTINIUM (AC) AS INDICATED BY THE COLOR CODE.

germanium in 1886. At that point, the value of the periodic table of elements was clear to all.

In the modern periodic table, Mendeleyev's rows and columns are reversed, but all the gaps are filled in. The order is not by atomic weight but by atomic number, an idea that did not come into being until 1911.

NEW QUESTIONS EMERGE

After Mendeleyev's discovery, the quest for new elements accelerated. Every element that was found fit into the periodic scheme that Mendeleyev first proposed. By the late nineteenth century, the periodic table of the elements was firmly established as the foundation of chemistry.

But like most great breakthroughs in science, the periodic table led to new questions. Probably the most significant were these: How many elements are there? What makes one element different from another? And why are the properties of the elements periodic? Those questions would not be answered during Mendeleyev's lifetime. They required a series of breakthroughs in the field of physics in the first three decades of the twentieth century.

And now in the twenty-first century, the quest for new elements continues in ways that Mendeleyev could never have imagined.

CONTROVERSY, CONSENSUS, AND THE NATURE OF MATTER

Any new idea in science needs to be tested before it is accepted. Many scientists challenged Mendeleyev's proposal that the properties of elements followed a periodic rule. Without an underlying basis for periodic properties, Mendeleyev could have just gotten lucky. Fortunately, by leaving gaps in the periodic table that he expected to be filled, Mendeleyev had made a prediction that could be tested by further research.

That is a common theme in science. Someone proposes a new idea, and someone else challenges it. As new evidence appears, the proposal is either strengthened or shown to be incorrect. Even if it is strengthened, there is often controversy over what the new data means. But eventually, a consensus or broad agreement emerges among scientists.

When a proposal achieves consensus, it is usually considered a rule, law, or theory.

In the case of the periodic table, scientists gradually recognized Mendeleyev's periodic arrangement as a rule that described the nature of the elements. The periodic table represented a "what" rule of matter, but it still was missing an underlying theory to provide the "how" or "why" of periodicity. Is there a fundamental property of matter that underlies the periodic behavior of the elements? The answer to that question is yes, but it would not be discovered until nearly twenty years after Mendeleyev's death.

FINDING MISSING ELEMENTS

In November 1875, French scientist Paul-Émile Lecoq de Boisbaudran (1838–1912) discovered one of Mendeleyev's missing elements. It was in the group of metals that contained aluminum. Mendeleyev had called it eka-aluminum, but Lecoq de Boisbaudran named it gallium. Some scientists said he chose the name because *gallus* is the Latin translation of *le coq* (the French word for "rooster"), but he insisted he named it after the Latin name for France, *Gallia*.

It was the kind of discovery that can give a new idea a boost in credibility. But more measurements of the newly found element had to be done. A year later Lecoq de Boisbaudran published a paper disputing Mendeleyev's prediction of gallium's density, or mass per unit of volume (such as grams per cubic centimeter). He measured the density of gallium as 4.7 times that of water, whereas Mendeleyev had predicted a ratio of 5.9.

When Mendeleyev read the paper, he suspected an error in the experiment and immediately wrote a long letter to Lecoq de Boisbaudran, asking him to redo it. At first Lecoq de Boisbaudran refused, saying that Mendeleyev, in fact, was the one who was wrong. A few months later, Lecoq de Boisbaudran conducted the experiment again and found that Mendeleyev was right. The specific gravity of gallium was in fact 5.9. Lecoq de Boisbaudran then published a new paper that supported Mendeleyev's periodic table of elements.

WHEN FRENCH SCIENTIST PAUL-ÉMILE LECOQ DE BOISBAUDRAN DISCOVERED GALLIUM IN 1875, HE PLACED IT IN THE PERIODIC TABLE GROUP THAT INCLUDES ALUMINUM. BUT HE SOON FOUND A PROBLEM. HE MEASURED GALLIUM'S DENSITY AS 4.7 TIMES AS MUCH AS WATER'S, DISAGREEING WITH MENDELEYEV'S PREDICTION OF 5.9. MENDELEYEV SUSPECTED AN ERROR IN PROCEDURE AND ASKED THE FRENCHMAN TO REPEAT THE EXPERIMENT. LECOQ DE BOISBAUDRAN DID SO, AND THIS TIME CONFIRMED MENDELEYEV'S PREDICTION. THE PERIODIC TABLE HAD SURVIVED AN IMPORTANT CHALLENGE.

RECOGNITION AND INFLUENCE

After Lecoq de Boisbaudran published his correction, Mendeleyev's career skyrocketed. Seemingly overnight, Mendeleyev was considered a genius. He was suddenly in demand throughout the scientific world to publish

and lecture about the periodic table of elements. He addressed numerous scientific conferences and spoke at ceremonies where universities offered him honorary degrees. Mendeleyev relished being at the top of his profession and was very proud that his work was widely recognized and celebrated.

Soon Mendeleyev's periodic table was being used all over the world. Like many great scientific discoveries, Mendeleyev's table pushed other chemists to further exploration. The gaps in the chart sparked a massive search for the missing elements.

THE DISCOVERY OF GALLIUM BROUGHT MENDELEYEV INTERNATIONAL ACCLAIM. HIS FAME ONLY GREW OVER THE NEXT TWENTY YEARS AS MORE NEW ELEMENTS WERE DISCOVERED TO MATCH HIS PREDICTIONS. HE BECAME A SOUGHT-AFTER SPEAKER AT MAJOR SCIENTIFIC CONFERENCES AND UNIVERSITY CEREMONIES, WHERE HE RECEIVED NUMEROUS HONORARY DEGREES.

Over the course of the next twenty years, Mendeleyev's periodic table of elements served as a guide for the discovery of many new elements. In 1879, Swedish scientist Lars Fredrik Nilson (1840–1899) discovered an element he called scandium (Sc) after the Scandinavian region of Europe that includes Sweden. Additional measurements showed that its properties matched those of Mendeleyev's predicted element eka-boron.

The next year, German scientist Clemens Winkler (1838–1904) discovered another new element, which he named germanium (Ge) after his home country. Its atomic weight matched perfectly with a missing piece in Mendeleyev's table. Many other elements followed, including the 1898 discovery of the radioactive elements radium (Ra) and polonium (Po) in uranium ore by Marie

AMONG THE MOST FAMOUS DISCOVERERS OF NEW ELEMENTS WERE MARIE AND PIERRE CURIE, SHOWN HERE WITH THEIR ELDER DAUGHTER, IRENE, IN 1904, THE YEAR AFTER THEY WON THE NOBEL PRIZE IN PHYSICS FOR THEIR WORK ON RADIOACTIVITY. MARIE CURIE LATER WON THE 1911 NOBEL PRIZE IN CHEMISTRY FOR THE DISCOVERY AND STUDY OF THE RADIOACTIVE ELEMENTS RADIUM AND POLONIUM IN URANIUM ORE. IRENE JOLIOT-CURIE WOULD LATER SHARE THE 1935 NOBEL PRIZE IN CHEMISTRY WITH HER HUSBAND, FRÉDÉRIC JOLIOT, FOR CREATING ARTIFICIAL RADIOACTIVE ELEMENTS.

Curie (1867–1934) and her husband, Pierre Curie (1859–1906). Polonium was named for Marie Curie's native country, Poland. Their careful chemical analysis contributed to a thorough understanding of naturally occurring radioactive elements.

Other discoveries of elements included a previously unrecognized group of elements known as noble gases because they do not form compounds naturally with any other elements. The first noble gas to be recognized was also the most common one, argon (Ar), in 1894. It makes up nearly 1 percent of Earth's atmosphere. Helium, the lightest of the noble gases, had actually been detected

THE DISCOVERY OF HELIUM

Helium is the only element in the periodic table that was discovered by an astronomer. While looking at the spectrum of the sun during a solar eclipse in 1868, French astronomer Pierre Jannsen (1824–1907) noticed a yellow spectral line that did not match any known element. Across the English Channel in England, Sir Joseph Norman Lockyer (1836–1920) was also observing the eclipse. He proposed that the spectral line was produced by a previously unknown element, which he named helium (He) after the Greek word for the sun, *helios*.

Helium was not detected on Earth until 1895, when Sir William Ramsay (1852–1916) detected it as a gas being emitted from the uranium ore cleveite.

by spectral analysis of sunlight during a solar eclipse in 1868, but it was not known to be a noble gas until it was detected on Earth in 1895. In 1902, Mendeleyev added the noble gases to his periodic table as a new group.

A PERSONAL SCANDAL

As Mendeleyev's career skyrocketed, his marriage to Feozva began to decay. The couple had never really been close, and now with Mendeleyev's tours and speaking engagements, they rarely saw each other. Eventually he fell in love with his niece's best friend, Anna Ivanova Popova. He divorced Feozva in January 1882 so that he could marry the much younger woman. The Russian Orthodox Church did not accept the divorce and considered Mendeleyev a bigamist. The public uproar that followed probably led to his failure to be elected into Russia's Academy of Science at the time.

A popular story, which according to the Russiapedia Web site may not be true, describes the public reaction to the divorce. If it was acceptable for Mendeleyev, why was it not acceptable for them? The Russian czar (monarch) reportedly said, "Mendeleev has two wives, yes, but I have only one Mendeleev." In any case, the divorce was clearly right for Mendeleyev. He and Anna had four children, and she was a very positive influence on his life in many ways. They loved each other deeply until his death on January 20 (according to the Julian calendar; February 2, according to the Gregorian calendar), 1907.

CHEMISTRY AND PHYSICS AFTER MENDELEYEV

Not surprisingly, Mendeleyev's periodic table of the elements led to many developments in the field of chemistry. But it also led to many insights in the field of physics. The discovery of radioactivity in 1896 provided one of the early signs that there was more to atoms than anyone had imagined. Until that time, scientists considered atoms to be the most basic particles of matter. But why were there so many different kinds of atoms, and what made the atoms of one element different from another?

The answers began to emerge in 1897, when British physicist J. J. Thomson (1856–1940) of Cambridge University discovered a tiny particle called the electron, which was much smaller and lighter than any atom. Within a few years, physicists and chemists realized that electrons were the parts of atoms responsible for valences and chemical reactions. They discovered that an electron carries a negative electrical charge. They also realized that as they went across the rows and down the columns of the periodic table, each atom had one more electron than the last. From that realization came the idea of atomic numbers.

Because an atom is electrically neutral, it had to have parts of it that carried a positive electrical charge.

And because electrons are so light, they realized that the positively charged parts, which they called protons, had to contain most of an atom's mass. But how did those parts fit together inside of an atom?

In 1909, physicists at Cambridge began to get their first clues. New Zealand-born Ernest Rutherford (1871–1937) and his students Hans Geiger (1882–1945) and Ernest Marsden (1889–1970) began experiments to find out. They directed positively charged alpha particles produced by radioactive atoms toward a thin sheet of metal foil. They knew that the alpha particles had enough energy to pass through the foil, but they expected them to collide with the atoms in the foil and change direction.

By detecting the alpha particles on the far side of the foil, they could tell how much their direction changed, and that would allow them to figure out the internal structure of the atoms. The results were astonishing. Most of the alpha particles went straight through without hitting anything, but some were diverted quite a lot. Some even bounced directly back. Rutherford described the result as "almost as incredible as if you had fired a 15-inch [38-centimeter] shell at a piece of tissue-paper and it came back and hit you."

Finally in 1911, Rutherford had figured out what the results were telling him. The positive charged portion of an atom that contained well more than 99 percent of its mass was concentrated in a region, which he called the

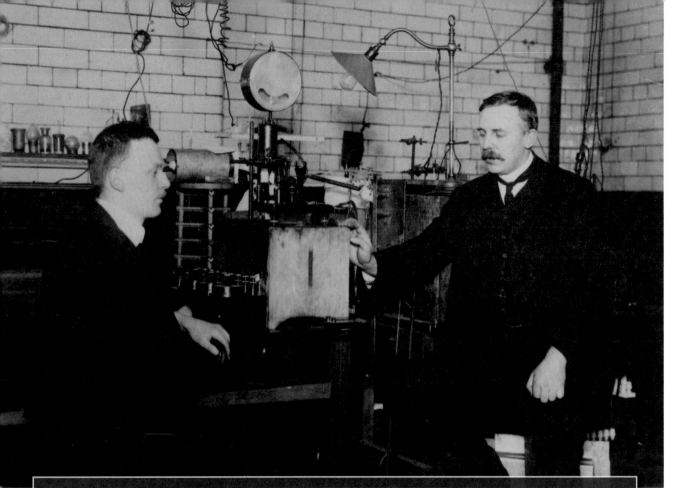

ERNEST RUTHERFORD, SHOWN HERE ON THE RIGHT, AND HIS STUDENTS HANS GEIGER (*LEFT*) AND ERNEST MARSDEN DISCOVERED THAT MOST OF AN ATOM'S MASS IS CONCENTRATED IN A VERY SMALL CENTRAL REGION CALLED ITS NUCLEUS. THE NUCLEUS CARRIES A POSITIVE ELECTRIC CHARGE EQUAL TO THE NEGATIVE CHARGE CARRIED BY THE ATOM'S ELECTRONS. THE ELECTRONS ARE RESPONSIBLE FOR CHEMICAL BONDING AND THE PERIODIC NATURE OF THE TABLE OF ELEMENTS.

nucleus, about one ten-thousandth the size of the atom. The rest of the atom is empty space except for the light-weight electrons.

Chemists realized that the bonds between different atoms in molecules were the result of sharing electrons, but the nucleus of each atom kept all of its protons. From that, they concluded that the atomic numbers in the periodic table corresponded to the number of protons in the atoms. That led to another puzzle because the atomic weights grew faster than the atomic numbers. Hydrogen had an atomic number of 1 and an atomic weight of 1. Helium had an atomic number of 2 and an atomic weight of about 4. Argon had an atomic number of 18 and an atomic weight of about 40. By the time they reached uranium, which had the largest known atomic number of 92 at the time, its atomic weight was about 238.

Rutherford theorized that the nuclei of atoms also contain other particles called neutrons that are about as massive as protons but carry no electric charge. That would mean that the atomic weight of an atom is close to the total number of its protons and neutrons. Not everyone accepted that theory until James Chadwick (1891–1974) detected neutrons in 1932.

NEW THEORIES OF PHYSICS

Rutherford's work marked the beginning of a new branch of physics called nuclear physics. Nuclear physics led to many advances, including an understanding of why certain nuclei are radioactive. It also led to techniques to make nuclei of new elements, called synthetic elements, that do not exist in nature. Including synthetic elements, the periodic table now goes well beyond uranium. All the heavier nuclei are radioactive and many of them decay very quickly into smaller nuclei. The largest as of 2013 is a noble gas with an atomic number of 118, whose discovery was first announced in 2006.

But why is the table periodic, and how do scientists know the chemical nature of those very heavy elements when they don't exist very long? The answer to both questions comes from another new theory of physics known as quantum mechanics or quantum theory. Every subatomic particle within an atom can be described by a set of numbers called quantum numbers. In 1925, Wolfgang Pauli (1900–1958) realized that the mathematics of quantum mechanics showed that no two protons, neutrons, or electrons in an atom

could have the same set of quantum numbers. For every electron, at least one of those numbers has to be different from any other electron.

Quantum theory predicts that certain numbers of electrons create what chemists call a closed shell. The electrons within that closed shell do not participate in chemical bonding. Only the electrons outside of the largest closed

IN 1925, PHYSICIST WOLFGANG PAULI DISCOVERED THAT THE NEW FIELD OF QUANTUM MECHANICS PROVIDED AN EXPLANATION OF WHY MENDELEYEV'S TABLE OF ELEMENTS IS PERIODIC. PAULI NOTED THAT EACH ELECTRON IN AN ATOM HAS ITS OWN UNIQUE SET OF FOUR QUANTUM NUMBERS, INCLUDING ONE CALLED THE PRINCIPAL QUANTUM NUMBER. THIS LED TO THE IDEA OF CLOSED SHELLS OF INNER ELECTRONS FOR EACH PRINCIPAL QUANTUM NUMBER, WITH A FEW OUTER ELECTRONS THAT DETERMINE THE ATOM'S PROPERTIES. TWO ATOMS WITH THE SAME NUMBER OF OUTER ELECTRONS HAVE SIMILAR PROPERTIES.

shell are involved in chemical reactions. Those outer electrons are also responsible for other properties of the elements, such as the way they form compounds or crystals, their electrical and magnetic properties, and the way they interact with light.

HONORING MENDELEYEV

Because the periodic table has been so important to chemistry and physics, Mendeleyev's name is one of the most honored in science. You can find it in many places, including on the modern periodic table. Here is a short list compiled by Moscow State University:

Element 101, mendelevium (Md)
Asteroid 2769 Mendeleev
Russian Scientific Ship named the *Dmitry Mendeleev*
Mendeleev Russian Chemical Society

Mendeleev Communications (A magazine published jointly by the Royal Society of Chemistry and the Russian Academy of Sciences)

Mendeleev Chemistry Journal (Abstracts—short summaries—in English)

D. Mendeleev University of Chemical Technology of Russia
Mendeleyevo (a town in the suburbs of Moscow)
Mendeleev Streets in Moscow and St. Petersburg
Mendeleev Monument in St. Petersburg

Elements in the same group in the periodic table all have the same arrangement of electrons outside of their largest closed shell. That is why their chemical properties and many of their physical properties are so similar.

MENDELEYEV'S LATER YEARS

When Mendeleyev produced the first periodic table in 1869, he was only at the midpoint of his life. He continued working actively throughout his remaining years and received many awards from various organizations. His textbook *Organic Chemistry* was awarded the Demidov Prize, one of the highest awards in Russia for educational texts. He revised it several times, and it was still one of the most popular chemistry texts in the world at the beginning of the twentieth century.

In 1882, Mendeleyev was honored with the Davy Medal "for an outstandingly important recent discovery in any branch of chemistry" by the Royal Society of London, England's oldest and most distinguished scientific association. In 1905, the society also gave Mendeleyev the Copley Medal, its highest award.

WORK AFTER THE UNIVERSITY

In 1890, Mendeleyev retired from his teaching position at the University of St. Petersburg. By then, he had been a part of the university for more than forty years, including the time he spent there as a student. According to the Woodrow Wilson National Fellowship Foundation, Mendeleyev's last lecture at the university included this reflection: "I have achieved an inner freedom. There is nothing in this world that I fear to say. No one nor anything can silence me. This is a good feeling. This is the feeling of a man. I want you to have this feeling too—it is my moral responsibility to help you achieve this inner freedom. I am an evolutionist of a peaceable type. Proceed in a logical and systematic manner."

After retiring from the university, Mendeleyev continued to work in other important ways. The Russian government hired him as the director of the Central Bureau of Weights and Measures, which monitors and assists scientific research and programs.

He also continued to publish his ideas. He had come a

MENDELEYEV CONTINUED STUDYING AND WRITING FOR MANY YEARS AFTER HIS OFFICIAL RETIREMENT FROM THE UNIVERSITY OF ST. PETERSBURG IN 1890. IN THIS 1903 PHOTOGRAPH, HE IS SITTING IN HIS OFFICE SURROUNDED BY BOOKS AND PAPERS, STILL AS PASSIONATE AS EVER ABOUT HIS FAVORITE SUBJECTS: CHEMISTRY, EDUCATION, AND THE HISTORY AND CULTURE OF HIS RUSSIAN HOMELAND.

long way since his first scientific article called "Chemical Analysis of a Sample from Finland," published at the early age of twenty. By the end of his career, Mendeleyev had written more than 250 science articles, which appeared in journals and science magazines. His two books, *Organic Chemistry* and *The Principles of Chemistry*, are still regarded as two of the most important science texts ever produced.

In 1906, only a year before his death, Mendeleyev published articles entitled "A Project for a School for Teachers" and "Toward Knowledge of Russia." Those highlighted two of his life's deepest passions: education and his homeland of Russia. He saw both sides of education, learning and conveying knowledge to others. Through his teaching, he shared his passion for science as a way to enlighten future generations. But teaching well requires the hard work of discovering new knowledge. Only through his own careful research and studying the research of others was he able to envision and construct the periodic table of elements, his gift to the world.

Because he actively shared that passion for science, Dmitry Mendeleyev remained a popular figure until his death. Fittingly, he died while sitting in his home, reading Jules Verne's *Journey to the North Pole*. Even at the very end of his life, he was envisioning new frontiers.

JANUARY 27/FEBRUARY 8, 1834 Dmitry Ivanovich Mendeleyev is born in Tobolsk, Siberia, Russia.

1847 Mendeleyev's grades in school improve, and his teachers proclaim him brilliant; his father dies.

1848–1850 The Mendeleyev family's glass factory burns down. The Mendeleyevs relocate to St. Petersburg, Russia; Dmitry enrolls in the Main Pedagogical Institute of St. Petersburg to study science.

1854 Mendeleyev publishes his first scientific article, called "Chemical Analysis of a Sample from Finland."

1859 Mendeleyev studies with Henri-Victor Regnault in Paris and then studies with Robert Bunsen and Gustav Kirchhoff in Germany.

1859–1861 Mendeleyev sets up a home laboratory and publishes many papers on his experiments with solutions.

1864–1865 Mendeleyev becomes professor of chemistry at the Technological Institute in St. Petersburg; in 1865, he receives his Ph.D. in chemistry.

1868 Mendeleyev finishes the first volume of his book *The Principles of Chemistry* and begins the second volume.

FEBRUARY 15–17, 1869 For three days, Mendeleyev works on how to group the elements. On the third day he dreams about the solution to the periodic table. He publishes his table in his second volume of *The Principles of Chemistry* later that year.

MARCH 6, 1869 A formal presentation of Mendeleyev's work is given to the Russian Chemical Society.

1870S Mendeleyev's table is challenged by the scientific community.

1880 A discovery by Clemens Winkler affirms Mendeleyev's periodic table.

DECEMBER 1882 Mendeleyev receives the Davy Medal from the Royal Society of London.

1890 Mendeleyev resigns from the University of St. Petersburg; Russian government appoints him director of the Central Bureau of Weights and Measures.

1894 Discovery of argon, the first noble gas to be added to the periodic table, though Mendeleyev did not add the noble gas group as a whole until 1902.

1895 Detection of helium on Earth, though its spectrum had been seen in sunlight in 1868.

1897 J. J. Thomson discovers the electron. Within a few years, scientists realized that the electrons in atoms are responsible for the chemical properties of elements.

1898 Marie and Pierre Curie discover the radioactive elements radium and polonium, beginning a detailed investigation of the chemistry of radioactive materials.

1905 Mendeleyev receives the Copley Medal from the Royal Society of London.

JANUARY 20/FEBRUARY 2, 1907 Mendeleyev dies at his home.

1909–1911 Rutherford discovers that most of an atom's mass is concentrated in a tiny positively charged nucleus.

1925 Pauli discovers the exclusion principle in quantum mechanics, which explains why the properties of elements are periodic.

2006 Element number 118 is first announced. As of 2013, it remains the largest known atom.

2011 The super-heavy elements 114 and 116 are officially recognized by the International Union of Pure and Applied Chemistry and are named in honor of the U.S. and Russian institutions where they were discovered. Element 114 is named flerovium (Fl) and element 116 is named livermorium (Lv).

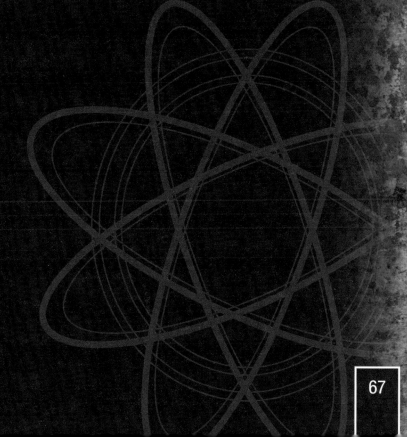

GLOSSARY

ALCHEMY A predecessor to chemistry, in which people tried to make certain substances out of others, often by heating things together. Most often, alchemists were searching for ways to turn less valuable metals into gold.

ATOM The smallest piece of matter that has the properties of a particular element.

ATOMIC NUMBER A number that gives the position of an element on the periodic table and is the number of protons it contains.

ATOMIC WEIGHT The weight of a single atom measured in units that are one-twelfth of the weight of an atom of carbon that has six protons and six neutrons in its nucleus.

CHEMISTRY The study of matter and the changes and interactions between different types of that matter.

COMPOUND A substance formed by two or more elements.

ELECTRON A tiny particle that carries the negative electrical charge in an atom and is responsible for the way that atom behaves chemically.

ELEMENT A substance that cannot be broken down into simpler substances through ordinary chemistry.

MASS The measure of the amount of matter.

MATTER Anything that takes up space and has mass.

MOLECULE A set of atoms joined together in a particular way, making it the smallest piece of matter that has the properties of a particular compound.

NEUTRON A subatomic particle within the nucleus of an atom that carries no electric charge.

NOBLE GAS An element whose atoms have no electrons outside of their closed shells and thus does not combine naturally with any other atom; the naturally occurring noble gases include helium, neon, argon, xenon, and radon.

NUCLEUS The tiny central portion of an atom that contains its protons and neutrons and almost all of its mass.

PHYSICS The study of matter, motion, energy, and force.

PROTON A subatomic particle within the nucleus of an atom that carries a positive electric charge.

QUANTUM NUMBER One of a set of four numbers that describes a subatomic particle within an atom. No two electrons in an atom can have the same set of quantum numbers.

SOLUTION Two or more substances that are mixed together.

VALENCE A positive or negative number that can be used to calculate the way atoms of different elements combine chemically to form molecules, which must have a net valence of zero.

WEIGHT The force that gravity exerts on a mass.

FOR MORE INFORMATION

American Association for the Advancement of Science (AAAS)

1200 New York Avenue NW

Washington, DC 20005

(202) 326-6400

Web site: http://www.aaas.org

Founded in 1848 and often called "Triple A-S," the AAAS is an international nonprofit organization dedicated to advancing science around the world. AAAS publishes the journal *Science*, one of the world's most widely read and respected sources of research reports in many fields.

American Chemical Society (ACS)

1155 Sixteenth Street NW

Washington, DC 20036

(800) 227-5558

Web site: http://portal.acs.org

According to its Web site, the ACS was chartered by the U.S. Congress and is the world's largest scientific society. It is one of the world's leading sources of trustworthy and accurate scientific information. Its stated mission is "to advance the broader chemistry enterprise and its practitioners for the benefit of Earth and its people."

American Institute of Physics (AIP)

1 Physics Ellipse

College Park, MD 20740-3843

(301) 209-3100

Web site: http://www.aip.org

The AIP is the umbrella organization for many different professional societies of physical scientists.
It publishes numerous journals for scientists and
magazines for educators, the public, and students
interested in careers in physics. Its Center for the
History of Physics contains a library and archive
of historical books and photographs.

Chemical Institute of Canada (CIC)

130 Slater Street, Suite 550

Ottawa, ON K1P 6E2

Canada

(613) 232-6252

Web site: http://www.cheminst.ca

The CIC is the national umbrella organization that includes the Canadian Society for Chemistry (CSC), the
Canadian Society for Chemical Engineering (CSChE),
and the Canadian Society for Chemical Technology
(CSCT). According to its Web site, its vision is that
chemistry is central to the well-being of society.

Dmitry Mendeleyev Museum and Archive of Saint Petersburg State University

Universitetskaya naberezhnaya, d. 7/9

St. Petersburg 199034

Russia

Web site: http://www.eng.spbu.ru/university/culture/ museums/mendeleev

According to the Web site Chemheritage.org, in a review of the museum (http://www.chemheritage.org/ discover/media/magazine/articles/26-2-mendeleev -at-home.aspx), the collection includes artifacts from Mendeleyev's life and work.

WEB SITES

Due to the changing nature of Internet links, Rosen Publishing has developed an online list of Web sites related to the subject of this book. This site is updated regularly. Please use this link to access the list:

http://www.rosenlinks.com/RDSP/dmend

FOR FURTHER READING

Aldersey-Williams, Hugh. *Periodic Tales: A Cultural History of the Elements from Arsenic to Zinc.* New York, NY: Ecco, 2011.

Belval, Brian. *The Carbon Elements* (Understanding the Elements of the Periodic Table). New York, NY: Rosen Publishing, 2010.

Bortz, Fred. *The Library of Subatomic Particles.* New York, NY: Rosen Publishing, 2004.

Bortz, Fred. *Physics: Decade by Decade* (Twentieth Century Science). New York, NY: Facts On File, 2007.

Brent, Lynnette. *Elements and Compounds.* New York, NY: Crabtree Publishing, 2009.

Bryson, Bill. *A Really Short History of Nearly Everything.* New York, NY: Delacorte Press, 2009.

Buckley, Don. *Introduction to Chemistry.* Boston, MA: Pearson Education, 2010.

Christianson, Scott. *100 Diagrams That Changed the World: From the Earliest Cave Paintings to the Innovation of the iPod.* New York, NY: Plume, 2012.

Dingle, Adrian. *How to Make a Universe with 92 Ingredients: An Electrifying Guide to the Elements.* New York, NY: Scholastic, 2011.

Green, Dan. *The Elements.* New York, NY: Scholastic, 2012.

Jackson, Tom. *Introducing the Periodic Table.* New York, NY: Crabtree Publishing, 2013.

La Bella, Laura. *The Oxygen Elements.* (Understanding the Elements of the Periodic Table). New York, NY: Rosen Publishing, 2010.

Lagerkvist, Ulf, and Erling Norrby. *The Periodic Table and a Missed Nobel Prize.* Hackensack, NJ: World Scientific, 2012.

Mullins, Matt. *The Elements.* New York, NY: Children's Press, 2012.

Padilla, Michael J. *Chemical Building Blocks.* Boston, MA: Pearson Education, 2009.

Robinson, Andrew, ed. *The Scientists: An Epic of Discovery.* New York, NY: Thames and Hudson, 2012.

Rogers, Kara. *The 100 Most Influential Scientists of All Time.* New York, NY: Britannica Educational Publishing, 2010.

Saunders, Nigel. *Who Invented the Periodic Table* (Breakthroughs in Science and Technology). Mankato, MN: Arcturus Publishing, 2010.

Scerri, Eric R. *The Periodic Table: A Very Short Introduction.* New York, NY: Oxford University Press, 2011.

Silverstein, Alvin, Virginia B. Silverstein, and Laura Silverstein Nunn. *Matter.* Minneapolis, MN: Twenty-First Century Books, 2009.

World Book, Inc. *Learning About Matter.* Chicago, IL: World Book, Inc., 2012.

BIBLIOGRAPHY

Allrefer.com Reference. "Transformation of Russia in the Nineteenth Century." Retrieved March 14, 2013 (http://reference.allrefer.com/country-guide-study/russia/russia23.html).

American Institute of Physics. "The Periodic Table of Elements." Retrieved March 14, 2013 (http://www.aip.org/history/curie/periodic.htm).

Andrew Rader Studios. "Rader's Chem4Kids." Retrieved March 14, 2013 (http://www.chem4kids.com).

Babaev, Eugene V. "Dmitriy Mendeleev Online." Moscow State University. Retrieved March 14, 2013 (http://www.chem.msu.su/eng/misc/mendeleev/welcome.html).

Bucknell University. "Russia in the Nineteenth Century." Retrieved January 14, 2013 (http://www.departments.bucknell.edu/russian/fn9015/1800Russ.html).

Corrosion Doctors. "Dmitri Mendeleev (1834–1907)." Retrieved March 8, 2013 (http://www.corrosion-doctors.org/Biographies/MendeleevBio.htm).

Giroud, Vincent. "St. Petersburg: A Portrait of a Great City." Yale University. Retrieved March 14, 2013 (http://beinecke.library.yale.edu/programs-events/events/st-petersburg-portrait-great-city).

Heilman, Chris. "The Pictorial Periodic Table." Retrieved March 14, 2013 (http://dwb.unl.edu/teacher/nsf/c04/c04links/chemlab.pc.maricopa.edu/periodic/about.html).

RT.com, Russiapedia. "Prominent Russian: Dmitry Mendeleev." Retrieved March 14, 2013 (http://russiapedia.rt.com/prominent-russians/science-and-technology/dmitry-mendeleev).

Rumppe, Roger, and Michael E. Sixtus. "Ich bin Mendelejeff." Woodrow Wilson National Fellowship Foundation.Retrieved March 14, 2013 (http://www.woodrow.org/teachers/chemistry/institutes/1992/Mendeleev.html).

Strathern, Paul. *Mendeleyev's Dream: The Quest for the Elements.* New York, NY: Thomas Dunne Books, 2001.

Third Millennium Online. "The Periodic Table." Retrieved March 8, 2013 (http://www.3rd1000.com/history/periodic.htm).

World Health Net. "Dreams Essential for Processing Memories." Science 2001; 294:1,052–1,057. Retrieved March 3, 2004 (http://www.worldhealth.net/p/237,268.html).

Zephyrus Interactive Education on the Web. "Dimitri Ivanovitch Mendeleev." Retrieved January 25, 2004 (http://www.zephyrus.co.uk/dimitrimendeleev.html).

INDEX

ABOUT THE AUTHOR

After earning his Ph.D. at Carnegie Mellon University in 1971, physicist Fred Bortz set off on an interesting and varied career in teaching and research before turning to full-time writing in 1996. From 1979 to 1994, he was on staff at Carnegie Mellon.

His books, now numbering nearly thirty, have won awards, including the American Institute of Physics Science Writing Award, and recognition on several best books lists.

Known on the Internet as the smiling, bowtie-wearing "Dr. Fred," he welcomes inquisitive visitors to his Web site at www.fredbortz.com.

PHOTO CREDITS

Cover (portrait), pp. 5, 21, 33, 36, 38, 45, 58, 63 Science & Society Picture Library/Getty Images: cover (periodic table) © iStockphoto.com/Jennifer McCormick; p. 9 Bruno Morandi/ The Image Bank/Getty Images; p. 11 Branger/Roger Viollet/ Getty Images; p. 15 Hulton Archive/Getty Images; p. 16 iStockphoto/Thinkstock.com; pp. 19, 30, 52, 53 Universal Images Group/Getty Images; p. 27 Mondadori/Getty Images; p. 34 The New York Public Library/Art Resource, NY; p. 42 Tahara Anderson; p. 47 Hemera/Thinkstock.com; p. 51 The Edgar Fahs Smith Memorial Collection, Special Collections Center, University of Pennsylvania Library; p. 60 Photograph by Samuel Goudsmit, courtesy AIP Emilio Segre Visual Archives, Goudsmit Collection; cover and interior pages (textured background) © iStockphoto.com/Perry Kroll, (atom illustrations) © iStockphoto.com/suprun.

Designer: Nicole Russo: Editor: Kathy Kuhtz Campbell; Photo Researcher: Amy Feinberg